THIS TEST CAN LEAD TO BETTER HEALTH

Below is the first part of the famous Bellew Test for determining your probable degree of arteriosclerosis. If you score 10 points or more in this section, take the entire five-part test you will find in the book (pages 35 and 36). It could be the most important thing you can do this year for your health.

The Bellew Test—Part I
History

1. Born after 1940 and urban raised . . . . . (5) _____
2. Formula (bottle) fed in infancy . . . . (5) _____
3. Processed baby foods (can, jar or cereal) (5) _____
4. Aging prematurely . . . . (5) _____
5. Overweight, present or past,
   except briefly . . . . (5) _____
6. Elevated blood pressure . . . . (5) _____
7. History of calorie dieting . . . . (5) _____
8. Exercising to reduce . . . . (5) _____
9. Family history of heart disease or
   diabetes . . . . (5) _____
10. Use of reducing pills at present . . . . (5) _____
    TOTAL . . . . _____

Introduction to Lecithin

Howard E. Hill

PYRAMID BOOKS NEW YORK

The Bellew Test was reprinted with permission from Let's LIVE Magazine, *Los Angeles, California, August 1971 issue.*

INTRODUCTION TO LECITHIN
A PYRAMID BOOK

Pyramid edition published October 1976

ISBN: 0-515-04219-6

Library of Congress Catalog Card Number: 72-186915

Printed in the United States of America

Pyramid Books are published by Pyramid Publications (Harcourt Brace Jovanovich, Inc.). Its trademarks, consisting of the word "Pyramid" and the portrayal of a pyramid, are registered in the United States Patent Office.

PYRAMID PUBLICATIONS
(Harcourt Brace Jovanovich, Inc.)
757 Third Avenue, New York, N.Y. 10017 U.S.A.

Dedicated to the memory of my daughter, Marilyn Ruth, who would be here today, had I known of Lecithin.

Contents

Introduction
What this Book Is All About

The extent of needless human suffering in this world has now reached astronomical proportions. And the truly terrible thing is that it *is* needless suffering.

Insane asylums, hospitals, prisons, and drug addiction centers are mushrooming. All these barbaric institutions are overflowing with victims and more facilities are being demanded. If you have ever been in one or if you know of someone who has become enmeshed in the toils of one of these outdated, hopelessly inadequate, and cruel institutions, I do not have to tell you how barbaric they are.

As we plow into the seventies, many more will have to be built to hold the growing numbers of victims of our society.

Each man, woman and child occupying a bed in any of these houses of horror has suffered and will suffer more because of our indifference.

TECHNOLOGY HAS CREATED PROBLEMS

Those sufferers are where they are because we have sought cures and palliatives and not causes. Our technology produces the causes of disease, illness, suffering and death. It breaks down our resistance through increasing daily stress; higher noise levels; pollution; foods leached of their nutritive properties and filled with preservatives and poisons for the benefit of the manufacturers, packers, shippers, and storage businesses; countless inducements to use drugs of all kinds, from caffeine to nicotine to booze to the hard drugs, until the victims are brought screaming off the streets by the hundreds of thousands every year.

IT IS UP TO US TO ACT

We know these things. And we have only scratched the surface.

What are we doing about it? Do we correct the causes? Not on your life! (Literally!) We continue to pack people into the institutions and to treat the symptoms, with drugs whose side effects are often horrendous and always throw the body systems off balance.

Do we use the knowledge of preventive measures that we have? Do we urge the curative and preventive forces of good nutrition provided by nature in foods and vegetable products that are available to us? We do far too rarely. Too often we ridicule them. Or we shrug and ask what can be done.

We pay for that indifference.

Countless lives are ruined. Countless families are destroyed. It is no small price to pay.

Must it reach the point that we hear persons in our own families sobbing in misery or screaming with pain, while we helplessly watch them gasping for breath through seemingly endless nights, before we will realize that our national way of disease production (as opposed to health production) is an insane, ruthless, and cruel way of life?

OUR CULTURE PROMOTES ILLNESS

One has only to look at the advertising on the billboards, in magazines and newspapers, and on television. Do you find the major advertisers trying to get you to buy foods and goods that are healthful for you? Few, if any.

They are too busy urging upon you alcoholic beverages to poison your body and addle your brain; tooth-rotting soft drinks loaded with empty calories and stimulants; processed foods; candies; huge gas-eating killer cars to slaughter you on the highways and pollute the air you breathe; coffee to load you with caffeine and make you nervous; soaps to pollute our streams and lakes with; and cigarettes to give you emphysema and lung cancer.

Worst of all, when you go to the man who should help you when all of these have destroyed your resistance and you have fallen ill, you are often handed inadequate prescriptions. You are given help that is *no* help.

What of the future? Think of the tired mothers

trying to make do on tight budgets who are forced to serve their children foods out of which all the best nutrient properties have been processed. Thus the children, the next generation, are set up to become the next victims. The vicious cycle perpetuates itself.

Should we forgive? Should we ignore it? The fact remains that human beings are being made to suffer needlessly by our very indifference and inaction. We are guilty of the sin of omission.

WE SHOULD FACE THE TRUTH

It is time to meet the problem head on and face the reality—fostering disease in human lives is the most brutal of all possible savageries. And that is just what our society has been doing. We frown on nations who would practice germ warfare. We must not let ourselves be guilty of practicing sickness warfare on ourselves and our children. Now is the time to wake up and do everything in our power to stop this insanity.

THERE IS HELP—AND HOPE

Substantial help is not far away and has been quietly developing for a long time.

Since the turn of the century, many miracle substances have been discovered in plants and other products of nature.

Perhaps one of the greatest of these is lecithin, a hidden resource of the lowly soybean.

ACCEPTANCE HAS BEEN SLOW

In the beginning, the distribution of soybeans and their by-products was limited mostly to progressive health food outlets, but as the fervor caught on, canners were packing the cooked beans, bakers were using it in bread, candy-makers were stabilizing chocolate with lecithin and health food stores throughout the land were selling millions of bottles of the wafers each year.

The demand is growing steadily, especially as the health benefits of lecithin and other soybean products become better known.

TRUTH IS EMERGING

So many conflicting theories are being expounded (for example some researchers suggest that the real benefit of using lecithin comes from the choline which is one of its constituents) and so many exaggerated claims have been made for lecithin that it is hard to separate the realities from wishful thinking and promotional exuberance. But the truth is now emerging that persons who suffer from heart trouble, arteriosclerosis, liver and kidney malfunctions, and other associated ills have received help—almost miraculously, in fact—simply by consuming one or two nut-flavored lecithin wafers every day.

That alone should make lecithin worthy of our considerate attention.

One:

Lecithin Brings Startling Results

The theorists can theorize and the researchers can formulate experiments, but what really counts is the empirical test: Does it work?

RETIRED DOCTOR OF PHARMACOLOGY PRAISES LECITHIN

Susan Chapman, age 86, a retired registered nurse, former head of one of the leading schools for nurses in the United States and a Doctor of Pharmacology, praises lecithin. "It is," she declares, "one of the really great natural substances available to mankind."

"When I first heard about the benefits attributed to lecithin," she went on to say, "I will admit that I was somewhat skeptical, but about a year ago

I decided to give it a try. For me, the results have been unusually good. Three major operations had left me underweight and lacking in energy. Within a comparatively few weeks I was back to my normal self. I had gained five pounds and I was feeling so much better that I began writing another book on the subject of nursing."

If you met Susan Chapman today, you would never guess her age. Her zest for living in the face of enormous physical difficulties is a real inspiration.

She continues taking classes at the Institute of Lifetime Learning in Long Beach, California, regularly attends a bookwriters' workshop, and takes a small but important part in both church and civic affairs.

PAINFUL LIMP DISAPPEARS, CHOLESTEROL COUNT DROPS

Elma Erwin, age 65, is a retired college professor. Two years ago Mrs. Erwin found she had an unusually high cholesterol count. Much too high to be safe according to her personal physician. On his recommendation, she began eating lecithin wafers "on a part-time basis," she said, "but the results were so good I now eat several wafers every day."

Mrs. Erwin went on to say, "My general health is better. My skin has more natural oil—and, of great importance to me—my nerves no longer give me problems. Being calm, I now face up to everyday situations much better."

When I first saw Mrs. Erwin, she was walking with a decided limp because of an extremely painful hip joint. This was diagnosed as a hard-to-relieve inflamed condition. Eating lecithin on a part-time basis failed to produce any noticeable results, but when she began eating the full recommended quantity, her condition began to improve rapidly. Her cholesterol count began to drop off steadily until it came within the limits of safety, and the last time I saw her walk into a room, I observed that she stepped along quite spryly with no evidences of discomfort.

I am moved to say that Mrs. Erwin's experience is not an isolated example. More and more people are discovering every day the benefits to be derived from eating high-quality lecithin.

— FROM BELOW PAR TO "LIKE A MILLION"

Audrey Marshal, age 42, of Santa Ana, California, is an excellent example of the beneficial effects of a strongly positive attitude combined with a regular intake of lecithin.

In a statement regarding her use of the granules, which contain known quantities of the components choline, inositol, and phosphorus, she declared, "About three years ago I found myself way below par physically and my thinking was not too clear. Since I am a novelist, it is quite necessary for me to be feeling my very best at all times in order to create salable copy."

She went on to say, "I will admit that my need for something to pep me up was influenced by an ad I read in a health magazine. Because the tired feelings described in the ad sounded so much like what I was experiencing, I went to my favorite health food store and bought my first package of lecithin. It is now three years since I started eating the product and I am delighted with the results. I feel better than I have felt in years and my thinking is no longer fuzzy—even after working at the typewriter for hours.

"In the beginning," Mrs. Marshal went on to explain, "I started to mix lecithin with my juices because it was easier to take that way. Now I sprinkle my daily ration on anything I eat. My purpose was to improve my general health. It worked. I now feel like a million and I am putting in a full day's work every day."

Mrs. Marshal is a personable and dynamic woman, full of enthusiasm for her work, and her experience is typical of what lecithin can do for people who have that run-down feeling. In more than one hundred interviews with persons of all ages, the reports were almost consistently the same.

CRIPPLED BY ARTERIOSCLEROSIS, HE WALKS AGAIN

Don McCallum, age 82, of Long Beach, California, had been enjoying his well-earned retirement after years of service as a telephone company executive up until two years ago. At that time he was advised by his doctor that he was afflicted with hardening of the arteries.

Since Mr. McCallum was familiar with the disease by name only, he was not prepared for what happened next.

"One morning," he said, "my legs were paining me so badly I could barely get out of bed." Naturally, he became quite concerned. He telephoned his doctor for an appointment and went through the regular routine of taking high-priced medication with little or no discernible results. The pain in his legs went from bad to worse, and the only way he could get around with any comfort was by using two canes.

After a few days of that arrangement, he stated, "I could hardly navigate at all."

In a chance encounter with a friend he was told about lecithin and all of the wonderful things it could do for one's health. Quite bluntly he told his friend, "I am more than dubious, but I will ask my doctor what he thinks."

Fortunately for Mr. McCallum, his physician was a man who was more interested in getting him well than he was in pontificating and dishing out prescriptions. "If you think it will do you any good, go ahead and try it," he said.

Mr. McCallum did, and the results were astonishing.

"Within four days," Mr. McCallum declared, "I was walking normally again. It was almost like a miracle. I now have only an occasional twinge of pain and I am feeling better by the day. I actually believe my arteriosclerosis has taken a beating."

SUCCESSFUL EDITOR
NEVER FELT BETTER

Yvonne Von Woerf, age 41, of Hollywood, California, a highly successful editorial director at a major west coast publishing firm, found herself in a bind with so many physical disabilities that she lost count. Tired of getting nowhere with both M.D.'s and Chiropractors, she turned to lecithin granules on the recommendation of a friend.

"I could be a little bit premature," she said, "claiming that all of my health problems have been solved, but the truth of the matter is I have never felt better in my life."

One situation that confronted Miss Von Woerf was high blood cholesterol, according to her doctor. "Hopefully," she said, "I began eating lecithin in the belief that this could be greatly reduced. I am looking forward to the results of my next blood test with more than ordinary interest. I will let you know if it is down from a count of 425, which the doctor considered high for a person of my weight and build. In other words, I was probably a fit subject for a cardiac arrest."

For many years, Miss Von Woerf took a very dim view of vitamins—and especially of all the propaganda going out about Vitamin E—but she is now a confirmed devotee of so-called health food products.

In explaining her attitude, Miss Von Woerf told me, "Excessively high prices, overselling and extravagant claims made for their products tended to cool me on taking any of the advertised items.

"To begin with," she stated, "I took to the lecithin idea with some misgivings, but, I reasoned, it surely couldn't hurt me—and it just might help." Judging from the results obtained so far, it was a wise gamble.

EDUCATIONAL MOTION PICTURE PRODUCER IS IMPRESSED

Agnes Goodring, age undisclosed, but plainly a person past 50, discovered one day that she was not handling location trips easily when producing pictures for educational purposes. Tired feelings were slowing her down so much that time spent on getting good pictures "in the can" was becoming a real problem. The cost factor was making it far too expensive for her to produce salable inventory.

When Mrs. Goodring was finally persuaded to go to a doctor, she was told that her blood pressure was much too high and that she very probably had arteriosclerosis and her general health was well below par.

After several tries with vitamins and tonics, she was still not as active as she felt she should be. Finally she resorted to self-diagnosis and began searching for an answer.

One of the so-called "health food nuts" told her about lecithin. Still somewhat of a skeptic, she decided to give this product a try. Results were slow in coming at first, but after several months she was back in full production stride ready to do a new picture.

As this is being written, Mrs. Goodring is planning a trip to the peak of Mt. Eisenhower on picture-taking venture that could easily extend into several months. Her blood pressure is now down to 140/80 and she says, "I feel like I could whip my weight in wildcats."

OVER ONE HUNDRED PEOPLE INTERVIEWED

In as fair a sampling as I could gather, I checked over one hundred persons selected at random who had decided to use lecithin in one form or another, hoping to correct a health problem. I discovered the following results.

Some nineteen percent of the persons questioned revealed a completely negative attitude toward themselves and their health and were quite critical about and dubious of the virtues of health foods, including lecithin, and any possible health benefits. Naturally this group was productive of very little information.

Fourteen percent of those interviewed were not too sure of improvement in their physical condition. Further questioning, however, revealed the fact that—almost without exception—this group had failed to have a blood count taken prior to including lecithin in the diet. And they had not taken the trouble to have either a blood pressure or cholesterol count taken even after as much as a full year of trying the product. As a consequence their information must be counted as very dubious. It was in this group that the greatest number of serious problems developed, according to our information, since we could not know anything of the uncooperative group.

The largest group, some sixty-seven percent of the persons interviewed, among whom were those who took the time to have reasonably frequent physical checkups made by their doctors to determine actual progress, were both highly responsive to interviews and strongly in favor of the benefits of eating lecithin. It seems likely that this group, being more careful to have reasonable checkups, is both more reliable in reporting (since they have accurate information) and more regular in following the regime.

On the scoreboard for the empirical test, then, lecithin gets highest marks in individual case histories and in broad samplings. Best of all, perhaps, by their own words, many who were suffering have experienced almost miraculous improvement. Lecithin does work.

Two:

Are the Claims
Made for Lecithin Valid?

It might be well, first, to know what lecithin is. It is a phosphatide, or phospholipid, a fatty compound found in all cellular organisms made up of phosphoric esters.

Lipids are our body fats. In association with proteins they are important components of our most important organs, such as the nerves and glands and the brain. Phosphorus plays an important role in our basic energy production in the Krebs cycle, in the high energy bonds of adenosine-triphosphate in the metabolism of every cell. This gives some inkling of the fundamental importance of phospholipids and phosphatides. Without phospholipids, we would have no brain, no glands, and no energy production.

Any good basic biology text or encyclopedia will provide further information on these important substances.

William C. Beaver and George B. Noland in their excellent *General Biology, the Science of Biology*, explain that when one of three fatty acids is replaced by choline and phosphoric acid in a molecule of neutral fat (such as palmitin) that lecithin is formed. Internally lecithin is a yellow-brown fatty substance, found in animal and plant tissues, composed of choline, phosphoric acid, two fatty acids and glycerol, according to Random House's *American College Dictionary*. Its name comes from the Greek lēkithos, which means egg yolk.

Lecithin is essential to cells in maintaining the surface tension of their membranes. Without it, presumably, every cell in our bodies would lose its structure and melt back into a primordial soup. Lecithin is also essential to the selective permeability of the membrane of every cell. This is vital for two reasons: 1) lecithin helps the cell control what comes into it and goes out, allowing nutrients in and expelling metabolic wastes, and excluding poisons, 2) the propagation of nerve impulses along the axons and dendrites of neurons (nerve cells) is dependent upon what appears to be a kind of sodium ion pump which in turn is dependent upon a delicately controlled selective permeability of the cell membrane.

All reliable sources agree that lecithin is essential to the proper functioning of every cell in every organ of the human body.

This is especially true, it would seem from the empirical evidence of people who have benefited, when it comes to supplying that extra-special zip and vitality to the working tissues of the liver and heart—and there is a strong indication that the lungs are greatly benefited, although this is still in the testing stages.

As this is being written, lecithin is steadily gaining a place in the formulation of remedies designed to cure diseases of the nervous system.

And it has now been established that lecithin slows down—often checks—the accumulation of fat in the liver. In addition to the foregoing advantage, it is also known that the use of lecithin tends to increase the absorption and assimilation of Vitamin A, as well as raising and sustaining the blood level of this important body builder.

Benefits Chart

IS LECITHIN A HEALTH BONANZA OR A GREATLY OVERRATED, OVERSOLD PRODUCT?

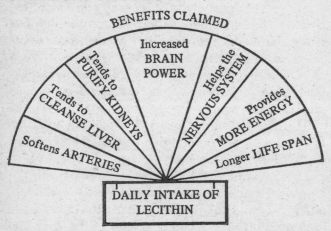

All evidence at this time seems to indicate that the "Benefits Claimed" chart shown here provides a fairly accurate picture of what LECITHIN can do for you.

Any additional experiences while taking LECITHIN should be recorded.

CAN LECITHIN CONTROL HEART DISEASE?

The big controversy raging at this time surrounds the use of lecithin in controlling heart disease, or lessening the harmful effects of arteriosclerosis. However, all evidence seems to point to the fact that the actual causation derives not from the whole product, but rather from a tiny part of the original yield of the soybean, known as *choline*. This vitally important part of the legume is often imitated synthetically in various forms, but true results are achieved only when organically grown soybeans are correctly processed and the genuine choline is removed for use in natural preparations.

Lack of lecithin, or choline, in the diet can produce enough kidney damage to cause serious health problems, according to research studies initiated in the University of Toronto, and later taken up by major universities in the United States. Without any question, soybeans should be included in the diet of all diabetics, declares Dr. H. W. Dietrich of Texas, because this food essential decreases the need for insulin. Tests reveal that it lessens the amount of sugar passed in the urine and that insulin requirements go down.

WHY YOU SHOULD KNOW

Many attempts have been made to correct the problem of arteriosclerosis. This effort has directed attention to the fatty content of foods. The reason for this is that during digestion fats are broken down into fatty acids. The words we see on labels: saturated (or polyunsaturated) explain the hydrogen content of these acids.

It has been stated that most fats are a combination of both types. Hard fats, normally solid at room temperature, are predominantly saturated. They include products that are hydrogenated, such as cooking fats, tallow, butter, most margarines, lard and fats that come from all meats. For the most part, unsaturated oils are liquids pressed from peanuts, corn and soybeans.

When a person is afflicted with arteriosclerosis, his blood fat is loaded with saturated fatty acids. On the other hand, when the blood fat of an individual is free from the disease there is a high percentage of unsaturated fatty acids present.

There are three established fatty acids. These are: arachidonic, linoleic and linolenic—all of which can be obtained from vegetable oils. Each is essential before cholesterol and saturated fats can be utilized by the body. When the diet furnishes a sufficient amount of linoleic acid, the other two essential acids can be synthesized from it, provided a wide variety of vitamins and minerals are eaten regularly. Unfortunately, several of these nutrients may be in short supply.

Obviously, many factors are involved, but when fats cannot be readily disposed of by the tissues, they tend to block the free flow of the blood. Because peanut, safflower, and soy oils are among the richest sources of arachidonic, linoleic, and linolenic acids, in the order given, competent authorities recommend mixing equal parts of all three for the best nutritional results. It is a balance in the diet that counts.

IMBALANCE: CHOLESTEROL— A TWO-EDGED SWORD

Preventive measures are desirable when imbalance occurs and the problem of high blood cholesterol prevail. Evidence indicates that the type of fat we include in our diet is very important. Of course, it is impossible to exclude all unsaturated fats from the diet, but balance should be our aim.

What is cholesterol?

Many people think that alcohol is always something that comes in a bottle. Well, not always. Cholesterol, a fat-like substance with the complex formula of $C_{27}H_{45}$ OH, is technically and chemically an alcohol, a solid alcohol. Its name comes from the Greek *chole* for gall or bile, combined with *ster(eos)* for solid and *-ol* a chemical suffix meaning that it is an alcohol. It was named that because it is a solid form of alcohol found in the gall bladder.

It is not inherently bad. Our bodies manufacture it because we need it, and we manufacture more of it when we are under stress and when our diets are faulty.

This solid white crystalline alcohol is a two-edged sword. Originally it is produced by the liver and then transformed into the several hormones used by the endocrine glands. It is changed into bile and it plays an essential role in the proper functioning of the brain, blood and glands.

How is it possible for such an important substance to also be a contributing factor in so many diseases? Arteriosclerosis, high blood pressure, even cataracts, have their beginning in too much cholesterol. Apparently, great harm is done when this substance becomes abnormal. When this body product is deposited in the veins and begins to slow or cut off free circulation of blood through all channels, disease takes hold.

This unfortunate consequence is usually caused by faulty fat metabolism. This process is a continuous operation by which all living cells undergo change. Our body is thrown off balance when short-changed by the foods we eat. This can happen through a faulty diet, genetic error, or stress.

Unfortunately, many persons in or out of hospitals, are encouraged to live on diets that actually increase the problem. For instance, many so-called low-cholesterol diets permit unlimited use of the five harmful "whites." These include manufactured refined sugar, white bread and refined cereals, salt, lard and treated milk, and white flour. Refined sugar, if not completely burned up in the system, is converted into saturated fat.

Normally, these diets forbid eggs and liver, while butter or hydrogenated margarine is permitted.

EXTRA BENEFITS FROM SOY OIL

Essential fatty acids are important to normal body functioning in a ratio of possibly one and two-tenths parts polyunsaturated fats to one part saturated. This is called the P/S Ratio. It is often recommended by physicians when treating heart patients. On the plus side many of the diets prescribed now recommend only safflower, soy or corn oil.

It is generally agreed that these oils are good, especially the ones that are cold pressed. Other cold pressed oils have some merit. The reason for this is that linoleic, linolenic and arachidonic fatty acids, as well as Vitamin E, are included.

Most competent dietitians agree that margarines made from these oils are not desirable. Labels should be studied for additives which are harmful. The so-called liquid margarines are widely acclaimed, even though they usually contain all too many chemicals. Better yet, locate your nearest reliable health food store and make your own spread. One recipe calls for one cup soy oil, a half-pound soft raw butter and four tablespoons of lecithin granules. Blend until smooth and keep well covered in your refrigerator.

A particularly nourishing spread high in essential fats as well as high-quality protein is made by mixing in a bowl one-third home ground peanut butter, one-third part soy meal and one-third part pumpkin seed meal. To this mixture, add two or three tablespoons of lecithin granules, high in choline and inositol.

The reason for including lecithin is that it is a great emulsifier of fats. It has been found that when there is a decrease in lecithin intake, conditions such as arteriosclerosis seem to thrive.

Adding supplementary lecithin granules to your

other foods is a good practice. It is known that the lecithin manufactured by the body comes from eating *unrefined* and *untreated* foods. It is apparent that natural oils aid both in lowering the cholesterol count and in helping to elevate blood lecithin. Without any doubt, hydrogenated fats or excessively fatty meats elevate cholesterol.

Other nutrients necessary for help in this cholesterol picture are choline and inositol which are part of the lecithin complex. Fresh vegetables and fruits are musts in the normal diet, as well as fresh vegetable and fruit juices.

WHEN LECITHIN BECOMES IMPORTANT

Lecithin, like cholesterol, is steadily and consistently produced by the liver. When it passes into the intestine, along with bile, it is absorbed into the blood. It is an aid to moving fats in the blood stream. It also helps body cells to get rid of excess fats and cholesterol. It is claimed that lecithin serves as structural material for every cell in our bodies. This is particularly true, according to nutritionists, for those cells of the brain and nerves.

Because lecithin is a powerful emulsifying agent, it is especially important in preventing and correcting the problem of arteriosclerosis. We know that blood is mostly water. Fats do not dissolve in blood. Consequently, when lecithin is present in normal amounts, it causes cholesterol and neutral fats to be divided into microscopic particles which can then be held in suspension. When this happens, it causes the minute globules to pass readily through arterial walls. Fats can then be utilized by our body tissues.

Hardening of the arteries is characterized by an increase of the blood cholesterol, with a parallel decline in lecithin. Research projects established the fact that simulated heart disease conditions could be produced by feeding the subject cholesterol. By the same token, the problem could be prevented simply by adding a small amount of lecithin to the diet.

It has now been determined that hardening of the arteries can be produced either by decreasing the intake of lecithin or increasing the cholesterol. When enough lecithin is included in the diet, the problem does not occur regardless of how much cholesterol is fed. This holds true even when arteriosclerosis is in an advanced stage. Normal body conditions are restored when lecithin is added to the diet.

PRESTO—FREE-FLOATING FAT DISAPPEARS

Where lecithin is used commercially as an emulsifying agent in the confectionary trade or in the baking industries, or in heavy manufacturing where oil must be broken into minute particles, it is an essential part of the process. In home cooking, lecithin has a decided advantage. For example, when you make gravies or salad dressings, two or three tablespoons of lecithin will make the separated or free-floating fat literally disappear.

This miracle emulsifying action is apparently the same whether outside or inside the body.

When lecithin from soy oil has been used in controlled tests, arteriosclerosis has invariably been decreased to a very marked extent.

Many physicians are now successfully reducing blood cholesterol with lecithin. Some doctors report that when four to six tablespoons of lecithin have been given daily to heart patients who had suffered attacks, the results were favorable. Although no other dietary changes were made, within six months the level of blood cholesterol dropped, in one case, from 325 to 196 milligrams.

Patients who eat lecithin invariably feel more energetic, have an increased capacity for work and are often relieved of body aches and pains. When the blood cholesterol has once been decreased, one or two tablespoons of lecithin daily have helped to keep the blood fats at normal levels. Even larger amounts can be taken over long periods with equally good results. Adding lecithin to the diet has also caused the pain of angina to disappear and has been especially beneficial to elderly persons.

If you fear you have arteriosclerosis in the early stages or if you know you have had high cholesterol levels, try the Bellew Test, and if your score is not good, see your doctor.

The Bellew Test

I History

1. Born after 1940 and urban raised _____(5) _____
2. Formula (bottle) fed in infancy _____(5) _____
3. Processed baby foods (can, jar or cereal)_____(5) _____
4. Aging prematurely_____(5) _____
5. Overweight, present or past, except briefly ____(5) _____
6. Elevated blood pressure_____(5) _____
7. History of calorie dieting_____(5) _____
8. Exercising to reduce_____(5) _____
9. Family history of heart disease or diabetes____(5) _____
10. Use of reducing pills at present_____(5) _____

TOTAL_____ _____

II Foods commonly eaten

1. Candy of any or all kinds _____(1) _____
2. Sugar of any kind for any use _____(1) _____
3. Cake, pie, pastry, doughnuts, etc. _____(1) _____
4. Ice creams, sherbets _____(1) _____
5. Soda pops with natural sugar _____(1) _____
6. Hotcakes, waffles, french toast, and syrup ____(1) _____
7. Jams and jellies _____(1) _____
8. Cookies, shortbread, brownies, etc._____(1) _____
9. Milkshakes, malts, freezes _____(1) _____
10. Cereal with sugar _____(1) _____
11. Processed fruit juices and nectars _____(1) _____
12. Artificial fruit juices or ades _____(1) _____
13. Cream substitutes, powder or liquid _____(1) _____
14. Sweet pickles and relish _____(1) _____
15. Potato chips, corn chips, other chips _____(1) _____
16. Popcorn, pretzels _____(1) _____
17. Crackers, all kinds_____(1) _____
18. Pastas, spaghetti, macaroni, noodles, grits ____(1) _____
19. Breads, all kinds and types, as rolls, etc. ____(1) _____
20. Liquid diet meals supposedly slimming _____(1) _____
21. Instant meals (liquid, powder, dry milk solids)(1) _____
22. Instant reconstituted milk from milk powder__(1) _____
23. Commercial yogurts, with or without fruit____(1) _____
24. Beers, ales, wines_____(1) _____
25. Processed lunch meats
    (sausages, cold cuts, etc.)_____(1) _____
26. Processed cheeses_____(1) _____
27. Processed tomatoes (purees, sauces, catsup)__(1) _____
28. Fast food meals, skimpy, unbalanced_____(1) _____
29. Pasteurized milks_____(1) _____
30. Onions, garlic, horseradish, mustard, spices____(1) _____

TOTAL_____ _____

III Life-diet age levels—as they apply to I and II

1. Infancy—0 to 1 year _____ (1) _____
2. Weaning—9 months to 3 years _____ (1) _____
3. Childhood—3 years to 10 years _____ (1) _____
4. Adolescence—10 years to 13 years ____ (1) _____
5. Youth—13 years to 21 years _____ (1) _____
6. Adulthood—21 years to date _____ (1) _____
7. Marriage—no matter the age _____ (1) _____

TOTAL _____ _____

IV Present age score—mark your age accordingly

1. 21 years to 31 years _____ (1) _____
2. 31 years to 41 years _____ (3) _____
3. 41 years to 51 years _____ (6) _____
4. 51 years to 61 years _____ (12) _____
5. 61 years to 71 years* _____ (24) _____

TOTAL _____ _____

*If you are over 71 you probably did not need to take this test in the first place.

V
Add up your total scores

I. _____
II. _____
III. _____
IV. _____

_____ TOTAL OVERALL SCORE

Divide the above score by the factor of 2, only if in addition to the excesses of sugar and carbohydrates and processed foods they were also accompanied by daily and monthly regular meals which contained whole raw milk, fresh meats, eggs and vegetables. Otherwise total overall score in V. will stand.

Degree of arteriosclerosis
(compare your score with below)

A. 0-10 _____ None
B. 10-20 _____ Minimal
C. 20-30 _____ Mild
D. 30-50 _____ Moderate
E. 50-70 _____ Moderately severe
F. 70-90 _____ Severe
G. 90 upward _____ Advanced

Three:

Oil for the Health of America

Research and development in the production of soya oil during recent years has resulted in tremendous progress. This is particularly true in delaying the reversion of this vegetable oil to its original beany flavor. Soya oil of edible quality is presently used extensively in the production of cooking and salad oils and in the manufacture of margarine.

Soy oil is extracted from the beans by three processes based on the principles of: 1) mechanical pressing—which depends on the use of heat, followed by crushing under extremely high pressures, 2) solvent extraction—involving the immersion of the oil-bearing seeds in a liquid which may be a petroleum product, and 3) a process known as cold pressing—a method preferred by people who follow the suggestions of health food lecturers. There is much to recommend in this practice.

FROM A FORAGE CROP TO BIG BUSINESS

Soybeans were formerly grown in the United States as a forage crop. Later developments brought the leguminous plant into use as a cover crop for the purpose of enriching the soil.

The plant's value as a source of food in the United States has grown in importance during recent years. Now America is one of the principal producing countries. Other nations which foster this crop include China, Manchuria, and Japan. With proper growing conditions, the plant grows to a height of three to four feet. Each plant grows from three to five clusters of hairy pods. Normally, each pod contains from two to four seeds each, depending upon the variety. The seeds mature in a diversity of color—including yellow, green, brown, or black. Each type varies considerably in size.

The oil content of the seeds also varies widely. This difference ranges from 11 to 25 percent of the ripened bean. Ordinarily, plants cultivated strictly for the extraction of oil, produce seeds that yield an average of between 15 and 20 percent oil. The yellow variety is generally preferred because of its better quality.

Harvesting of the beans takes place as soon as possible after the plant reaches maturity. The reason for this is to avoid damage from frost and fracture of the seed coats. Any injury to the ripened pods permits the entry of fungus with resultant deterioration of the oil and protein content. Damage to the crop also greatly reduces the quality of the lecithin intended for human nutrition.

SOY OIL—THE SOURCE OF LECITHIN

Crude soybean oil contains a mixture of phosphatides—better known as lecithin. The amount of usable products varies with the method by which the oil is removed from the seed. However, the normal yield ranges from 1.5 to 3 percent of good grade beans.

One of the most important properties of lecithin is its effectiveness in lowering the surface tension of watery solutions. The reason for this is that, as with the soap molecule, one side of the molecule is attracted to fat, while the other side prefers water. It is this unusual Jekyll-and-Hyde characteristic that serves to make lecithin a most powerful emulsifying agent. It is to this unique property that it owes its greatest usefulness in the human body where arteriosclerosis is a problem.

Commercial uses of lecithin include its application to pharmaceuticals and cosmetic preparations.

MARGARINE MAKERS LEAD THE WAY

One of the early uses of lecithin was in the manufacture of oleomargarine. The addition of a small amount of lecithin to oleomargarine in the mixing stage insures an even dispersion of water throughout the oil product.

A commercial miracle has come to pass. With

all odds against soybean oil, it has become a staple market item. Four important factors have served to crystallize this success: 1) a good product, 2) reasonable prices, 3) a merchandising program alert to consumer needs, and 4) a strong, continuing research program.

These tremendous gains were made in the face of great competition. It is well to note that edible or industrial fats and oils may be obtained from established animal sources such as hogs, cows, sheep and whales. In the plant world, cottonseed, linseed, corn, and peanuts. Literally oceans of edible and industrial oils are required to care for consumer demand. At the present time more than eighty pounds of oil per person are required to satisfy the yearly needs for home and industrial use.

THREE MAIN USES OF SOY OIL

The use of fats and oils falls normally into three main categories: 1) home cooking absorbs less than three-quarters of all production, 2) about 20 percent is used for soap-making, and 3) about ten percent goes to the drying industries such as the manufacture of paints and varnishes.

THE TRADE WAR IS ON

As a result of the inroads made in the field of edible fats and oils, a lively commercial war soon developed. It now appears that soybean oil is coming off a solid victor.

Producers of domestic fats and oils have witnessed broad revision in the agriculture of their respective regions since the end of World War II. Acreage formerly devoted to raising cotton has been turned to cultivating peanuts or other more profitable feed crops. Surplus corn acreage is now being planted in soybeans. Soybeans yield more oil per acre than corn when used to fatten hogs for the production of lard.

Soy oil will doubtless continue to grab a larger share of the market. Farmers will grow more soybeans. Manufacturers are going to find new uses for soy oil. Admittedly, the market for oil is immense, but so is the reserve supply of oils and fats required to fill the growing need.

SOYBEANS INEXPENSIVE TO GROW

Unlike cottonseed oil, soybean oil is obtained from a crop grown on mechanized farms and requires a minimum of costly human labor to harvest. This factor alone had lead to a relatively low selling price. Moreover, its stable low cost points to a continuing growth in its use as a basic industrial material. Production will doubtless be greatly expanded as the market needs grow.

Steady improvement in processing techniques and improved refining equipment, combined with a well-financed research program, have brought forth an excellent product suitable for use in the production of tasty mayonnaise and salad dressings.

Step-by-step, chemists and engineers have added improvements and standardized production methods. Users of edible items based on soy oil began consuming the output in increasing volume. This progressive step served to reap advantages in cost and quality. Other manufacturers soon fell into line—some from necessity and some in pursuit of the potential health food market.

About this time a curious thing happened. It was discovered that soy oil imparted to margarine a rich, yellow, butter-like color. To offset this intrusion into the sacrosanct dairy industry, restrictive laws were promptly pushed through both congress and state legislatures. Within the next few years, research proved that soy oil could be hydrogenated and used in margarine with admirable results—even though the product in that form was no longer helpful in reducing arteriosclerosis.

FROM HOG FAT TO VEGETABLE OIL

At this time about twenty edible oils are used for cooking. Each is highly competitive. Fluctuation in price is a constant trade situation. Today a manufacturer might be using cotton and peanut oils in his product. Next month market conditions might indicate that cotton and soy oils could help him to keep production costs down.

It was into this competitive market that the soybean had virtually to hammer its way to the top. However, the market is growing by the day.

It is well to note that all edible oils and fats in their pure state have about the same caloric value. Competition steps into the picture because of the interchangeability factor. Take shortening, for example. Several different oil types can be used alternately by blending various consistencies to achieve a combination that leads to the desired body. Results that are watched for are color, flavor, odor, stability, plasticity, creaming ability, and emulsifying power. The ultimate consumer is interested only in the end product—not in the type of oils used.

Four:

Commercial Uses
of Lecithin in Foods

What we know as commercial lecithin is a mixture of oil, lecithin and lecithin-like products. For example, typical lecithin deriving from soybeans may have 15 percent oil and 25 percent lecithin. Other by-products make up the remaining 60 percent described as cephalin and galactolipids.

When lecithin is extracted from soybean oil, it possesses both emulsifying and stabilizing properties useful in commerce. This apparently is not true when lecithin is extracted from egg yolk. The cephalin fraction of vegetable lecithin and the phospholipids derived from soybean oil have been found to possess excellent *antioxidant* properties. Separate portions of the cephalin molecule have been isolated and patents have been granted to researchers.

Following the solvent extraction of lecithin from

soybean oil, the other phosphatides are removed by washing. This can be accomplished with water or diluted citric acid. The remaining phosphatides produce about 30 percent of useable lecithin.

The resulting sediment produced by the centrifugal treatment of the washed oil is dried by a vacuum process at approximately 140°F. Bleaching can then be completed with hydrogen peroxide.

LECITHIN USED BY BAKERIES

Soybean lecithin is now widely used by progressive bakeries. The improvement in product can be summarized as follows:

(1) When the free water is emulsified with the shortening, it gives the dough a drier feel. Consequently, it is less sticky. This results in easier handling.

(2) The emulsifying property of lecithin causes a more uniform distribution of the shortening.

(3) Lecithin imposes a favorable influence on the manageable properties of gluten because it effectively lubricates the strands of gluten.

(4) There are many other beneficial results. For example, it produces better texture and grain, a longer shelf life, and a more tender crust.

LECITHIN USED BY CANDYMAKERS

The candy industry uses soybean lecithin extensively. Some of the many products include fillings, caramels, chocolates and ice-cream bars. Because of its emulsifying property, lecithin causes all vegetable and other butters to mix more thoroughly with the other candy ingredients. This results in materially reducing the use of the more expensive fats, such as creamery butter. Lower cost and better storage quality are thus accomplished. Lecithin also serves to reduce what is described as fat bloom on chocolates. This is caused when the oils used in manufacture of the confections move to the outside surface.

Because of its antioxidant properties, lecithin extracted from soybean oil is effective in stabilizing the oil against oxidation. Deterioration of flavor is thus avoided. In this use it is as effective as citric acid.

LECITHIN USED IN FLAVOR EXTENDERS

Commercial production of glutamic acid and monosodium glutamate depends upon the separation of glutamic acids from natural proteins. The product is usually of vegetable origin.

Glutamic acid was first isolated from protein by a German chemist name Ritthausen in 1866. This talented researcher introduced into protein chemistry the tentative means of acid hydrolysis. Although the method was forgotten for many years, the experimental work started by Ritthausen formed the basis for the manufacture of glutamate in the United States.

At this time it seems reasonable to assume that the contribution of glutamate as a food flavor extender had its origin in the oriental process used to make soy sauce from soybeans.

OUTSTANDING FAST-WHIPPING AGENT

Albumins have gained wide acceptance because of their ability to speed up the whipping process. This special quality helps to hold finely dispersed air so as to contribute to a better texture and greatly improved characteristics of foods in which the product is used. Soy albumin has gained increasing acceptance since 1940. It has proved desirable because of its whipping properties.

Albumin derived from soybeans will beat up any food lighter than will egg albumin. Also, it will not overbeat even when it is left too long in the beater. There is one drawback. It will not stand up as long after whipping as will egg albumin. Because of its advantages in cost it is successfully replacing egg albumin in many manufacturing practices, despite its limitations.

In the manufacturing process of extracting albumin, the procedure begins with soybean flakes. This basic material is the oil-free, low-fiber protein of the bean.

Soy albumin is, in fact, a product different from egg albumin, even though both are protein in nature and valuable as aerating agents. In the manufacture of candy, they have little in common:

(1) To begin with, soy albumin does not "beat down" as egg albumin does after the peak increase has been reached. In practice, soy albumin continues to gain in aeration and to "firm up" the acquired consistency with prolonged beating.

(2) Normally, soy albumin produces more aera-

tion than egg albumin on a per-pound basis within a given time.

(3) Another very important difference between egg albumin and soy albumin is that egg albumin will heat-coagulate. Soy albumin will not. For that one reason, egg albumin will make meringue. Soy albumin will not, even though the soy product is preferred in the production of marshmallows.

OUR NUMBER ONE CASH CROP

The soybean heads the list of cash crops grown in the United States. In 1970 approximately 43 million acres of these valuable legumes were harvested. At this time, most of the acreage devoted to soybeans is in Illinois, Missouri, Indiana, Ohio, Iowa, and Minnesota. South Carolina, North Carolina, Arkansas and Mississippi lead in soybean production in the South, accounting for less than 20 percent of the acreage.

Soybean plants are grown mainly for their beans, which are used to produce oil and meal for commercial use. Soy meal, for example is used almost exclusively for livestock and poultry feed in the United States. Only a small part is devoted to improving food and industrial products. Most of the oil consumed in the United States is used primarily in foods such as shortening and margarine.

Manufacturers of paints, varnishes, and other industrial products make good use of the remaining by-products.

Soybeans are also grown for the purpose of

plowing under to enrich the soil. The seeds are used for food in areas where they are plentiful.

SOYBEAN OIL USED IN OIL-SPILL TESTS

Soybean oil will be used by the Coast Guard in Southern California for an off-shore test in an attempt to perfect a new barrier device designed to trap accidental oil spills on the high seas. Fish like the flavor of soy oil far better than crude oil.

The tests are scheduled to be conducted in early spring off Point Conception. If the tests prove effective, it will be the first method to succeed in containing crude oil spilled in the open ocean.

The Coast Guard tested the device earlier in calm waters off Tampa, Florida. Six hundred barrels of soybean oil were dumped into the ocean—with favorable results.

The man-made barrier can be anchored in a U-shaped configuration downcurrent from a spill. In this manner the oil is trapped as it drifts into the open mouth of the barrier. The big advantage here is that when there is no appreciable current or if the oil is drifting toward the beach, the barrier can be towed in a U-shaped fashion to overtake the oil.

To complete the work of the crude oil catcher, removal equipment is being developed in a separate coordinated project. The Coast Guard has plans to have barriers located in strategic coastal ports around our coasts. Probably some of the devices will be near Coast Guard air stations so they may be transported quickly to the scene of an oil spill. The Coast Guard hopes to be able to deliver a packaged barrier to an oil spill scene in less than four hours.

The Coast Guard explains that soybean oil will be used in all tests because it does not endanger marine life. The oil is non-toxic, bio-degradable, and non-persistent. It can be consumed by fish without any ill effects.

THE CHEMISTRY OF NATURE

School children often dream of asking their teachers to explain the startling fact that when a red cow eats green grass, she will give white milk which, in turn, can be made into yellow butter and blue cheese. Many profoundly learned chemists would also like to know of an authoritative answer to that one. In return, some of them might offer their own version of another natural wonder.

Take a field of soybeans for example. Sheep flourish on a diet of soybeans as with hay. A suit of clothes can be made from sheep wool. Milk cows and beef cattle also do well when fed soybean silage. They also enjoy the protein concentrates made from soybean meal. Dairy cows will then produce plenty of milk. After a quantity of food values have been removed from the milk, whey is left. Casein is made from the whey. From the casein a protein fiber, very similar to wool, can be processed. A tailor can take a fabric made with the casein fiber and fashion a smart suit of clothes.

With the protein taken from soybeans, a chemist can produce a pure soybean protein fiber which can also be made into a suit of clothes. And why, he might ask, should one bother with sheep or cows when an acre of soybeans will yield a far greater supply of cloth than the sheep can produce?

Fiber made from soybeans is strictly a synthetic fiber. The soybean produces no large amount of fibrous material comparable to a cotton ball or a hemp stalk. First the protein has to be isolated from the rest of the bean. The sticky substance is subjected to treatment in a suitable solution of chemicals. It is then squirted through spinerettes into a special bath which serves to fix the fiber. It is then wound into skeins. The result is a mildly lustrous, loose and fluffy material, light tan to white in color. It feels soft and warm to the touch. It has a natural crimp and is highly resilient, not quite as strong as wool because moisture considerably lowers its tensile strength. Otherwise, it is equally good. When the yarn is half of wool, half of soybean fiber, the resulting material is said to make durable upholstery fabrics. It is also suitable for other purposes, including blankets.

Laboratories are now producing a transparent wrapping material or film produced from soybean protein. There are strong indications that it could become a competitor in the market place.

Five:

From Backstage to Center Stage

Shortly after the turn of the century, a backstage room in Seattle's old Grand Opera house, housed the real beginnings of most commercial uses of soybeans.

It was there that a young chemist by the name of Irving Fink Laucks who owned a small laboratory which performed chemical analysis for the trade, began his research and study of the soybean. Requests for his custom service normally kept his force busy full time, but there were odd hours available for practice projects.

In looking around for a promising enterprise that would serve to keep the staff occupied and that, if properly developed, would bring more business, he was indirectly led to study soybeans.

The Northwest's burgeoning plywood industry,

feeding upon the huge forests of Douglas fir, provided the first clue. Wood was available in great quantities and the mills had learned how to slice it thin enough to manufacture plywood. A glue was needed to hold the fragile pieces of veneer together. Starch and animal glues were plentiful, but they lost all of their adhesive qualities when exposed to moisture. In brief, they were not water resistant. Naturally enough, this imposed definite limits upon creating dependable markets for plywood products.

Laucks took a close look at the problem, examined the raw materials from which the adhesives they had been using were manufactured and the big quest was on. One substance after another was tested and discarded, but somehow he always came back to soybeans as a basic material. Remember, this was well before soybeans had become much of a farm product anywhere in the United States. Since he had been called upon to make an analysis of this foreign legume, Laucks knew that soybean meal from Manchuria was arriving in Seattle in small quantities to be sold as livestock feed. It was easy to buy the big cartwheels of soybean cake in the market place.

First, he tried grinding the cakes into a flour, hoping to make them into glue. No two batches turned out the same until he learned to select cartwheels of a quality suitable for his purpose. Now the difficult job of selling began. Getting the plywood manufacturers to adopt his product was a real challenge. The equipment in use was not adapted to spreading this new kind of adhesive.

After many frustrating trials, Laucks finally landed two customers, but it was still hard going. There was a simple reason why.

Laucks' soybean glue had an unfortunate characteristic: it was not tacky.

When buyers applied the traditional test for tackiness, the new glue failed. This involved taking a bit of wet glue in the palm of one hand. The buyer would pat it with the fingers of his other hand to determine if sticky strings would adhere to his fingers. This step usually lost him the sale. Soybean glue would not do that.

Men in the lumber mills described the new glue as bean soup and dismissed the product as of no value.

Use of soybean glue was also hindered by the fact that so few plywood mills knew how to use soybean glue. Laucks usually had to post his men in key production spots to help with the operation. "This was probably a good thing for our cause," he commented in later years. "It kept us from making too many mistakes too far from home."

AUTO MAKERS ENTER THE SCENE

About this time automobile makers began to buy plywood for many purposes. It was used first in floorboards and later as instrument boards and finally for running boards. The plywood industry was in for a shock. Running boards were falling off. They were not sufficiently resistant to water.

The industry was now ready to unite to seek a

solution to the problem for the first time. The first move was to appoint a committee to select, after exacting tests, a glue that would meet the demands of the car manufacturers. It was agreed in advance that all participating companies would install whatever equipment was necessary to use the glue that passed rigid tests. The tests were made in 1926 and 1927. It was a time of great stress for Laucks and his men. His company was in line to get all of the business or be wiped out. Near the end of 1927, Laucks and his valiant organization had all of the business on the Pacific Coast.

This new acceptance called for soybeans in great quantities. Laucks visited Eugene Funk in Illinois and within a short time entered the market as a substantial user of soybeans, in the form of oil meal. Later on he was able to take on Eastern plywood makers as customers.

By this time, Laucks was ready to adapt his soybean glue to new markets for adhesives. This included tie sizing materials used in clay products and whiting to finish high-grade paper. This project fizzled. Traits of the glue base were not appropriate. However, one great good did come out of the experiment: His men created the first washable wallpaper and the Laucks organization has been able to hold on to almost all of this market.

Laucks never let up in researching problems peculiar to the plywood industry. Because of this strong interest, Laucks eventually developed a process that used soybean glue to manufacture a plywood that was not only water resistant, but also waterproof.

LECITHIN ENTERS NEW FIELDS

As though to establish the many-sided virtues of the soybean, it was discovered that lecithin was a valuable antifoaming agent. A big test came when leaping and exploding flames of high-test gasoline threatened to destroy an aircraft carrier. Hundreds of fine young American boys were about to meet a tragic death. Soybean products were ready for this terrible emergency. In the capable hands of the carrier's fire-fighting crews, the fires were extinguished in almost exactly one minute.

The preparation that quelled the fire on the huge ship in just sixty seconds was fire-foam. The basis of fire-foam is a derivative of soybean meal. Fire-foam, when sprayed under pressure in a mixture of 94% water with 6% of the soy product, is able literally to cover the flames with a blanket of tough bubbles six to eight inches deep. This foam excludes oxygen. With the dangerous gases sealed in, the fire has nothing to feed upon and it goes out.

WAR NEEDS DELAYED MANY USES

The critical need for oils and fats during the second World War, led to severe restrictions upon other than food uses of soybean oil. This restraint probably delayed many valuable technical applications. Since the war, a few of the end products in which you, as a consumer, would be likely to note the use of soybean oil are listed by one authority as: "paint, lacquer, enamel, resins, putty, caulking compounds, oilcloth, linoleum, shade cloth, oiled clothing, leather, wallboard, core oils, oiled parchment, sulphonated oils, lubricating greases, printing ink, disinfectants, insecticides, soap and waterproof cement."

During the past twenty-five years, constant improvements have been made in all refining processes. Every effort is being made to improve its qualities for each special use. Soy oil processors work hard to make their product not only desirable, but greatly superior and whenever possible, completely indispensable.

Industrially, the lowly legume is a one-bean band. It possesses an almost unbelievable variety of uses. According to Whitney H. Eastman, president of the Vegetable Oil and Protein Division of General Mills, Inc., near the end of the war General Mills Research Laboratories had counted one hundred and eighty-one soybean products: "All of which," he added, "might have numerous variations as to type and properties." The list by now easily exceeds five hundred new uses.

THE FAMOUS EXHIBIT CAR

One far-sighted general agricultural agent, a Mr. East, of the Pennsylvania Railroad in the late thirties, conceived the idea of an education dramatization of progress made with soybean production. He persuaded the railroad to build a soybean exhibit car to cover all the lines served by Penn.

The car ran on wheels which were cast with the aid of soy protein and oil used in the foundry cores. Inside and out, from top to bottom the car was protected with paint and varnish made from soybean oil. The interior finish was made from plywood held together with soybean glue. The exhibit car was packed with materials made from soybeans. The many colorful exhibits included paints, wallpaper, soap, fly spray, automobile parts, linoleum, explosives and livestock feeds, as well as foods for humans.

The exhibit car paid off handsomely for the railroad. By encouraging farmers to grow the crop during subsequent years, the railroad hauled more soybeans as freight. This resulted in more goods going back to the farm towns. Later on, other railroads took the car on loan to show the exhibit on their own lines. During nearly eighteen thousand miles of travel through eighteen states, close to two hundred thousand persons walked through the exhibit car.

This outstanding promotional idea was able to advance soybeans as a farm crop. This was highly successful because of one fundamental fact: The United States has never been able to produce all of the high-grade protein that customers want. Vegetable oil has always been in short supply. The rest of the world shares this problem. Soybeans seem to be the answer. It is an efficient and fertile source of both of these basic commodities.

Six:

Do You Need Choline?

Choline is a basic constituent of lecithin. The emulsified phospholipid is a combination of fatty acids and phosphorus. In the average diet, lecithin is found only in egg yolk. Lecithin is an important part of soybeans. Most relevant to our present study is this astonishing fact: Human mother's milk contains lecithin. Cow's milk does not.

Apparently, your whole family needs choline. From toddler to senior citizen, it should be included in your diet. It is part of the lecithin complex. The need for this valuable food additive begins at birth. The American Journal of Public Health, March, 1966, published a paper by Dr. W. S. Hartroft, M.D., Ph.D., in which he reported that a lack of choline was found to set infants on the path to high blood pressure.

THE NEED FOR CHOLINE IS ALWAYS PRESENT

From the foregoing, it seems to be obvious that the infant whose mother does not nurse him, but feeds him on a formula based on cow's milk, has a greatly increased chance of building up a deficiency in choline—right at a time when he should be acquiring this vitally needed nutrient.

Another situation that is quite alarming: it is claimed that once there has been even a brief period of choline deficiency during infancy, adding choline to the diet in later life will not reverse the tendency toward high blood pressure. This finding has now been confirmed in a number of laboratory tests conducted by competent researchers. Consequently, it is suggested that the best way to avoid a choline deficiency in your child is to nurse him from birth. When the baby is weaned, you should include lecithin with the balanced factors of choline and inositol.

HOW MUCH IS ENOUGH?

You should try to eat at least 150 milligrams of choline daily. This amount can be found in brewer's yeast or desiccated liver tablets. Each gram of desiccated liver contains about 10 milligrams of choline, while each gram of brewer's yeast contains 24 to 36 milligrams.

A VITAL ORGAN PROTECTION

It is claimed that this little-known food additive, choline, offers an important but valuable safeguard to the nerves, liver, and kidneys, provided you do not extend its protective ability too far.

It is just plain common sense to know that even with abundant amounts of choline in the diet, you cannot afford to eat rich pastries, butter, ice cream or any other excessively fatty foods—especially foods fried in hydrogenated fats.

Without any question, you should consume some fat, for the essential fatty acids, but your intake should be in the form of vegetable oil, made from peanuts, sunflower seeds, and naturally emulsified fat such as is provided in soybeans. Quite obviously, you should not let careless eating habits destroy the protection that choline can supply.

TOO LITTLE IS KNOWN

Unfortunately, this little-known member of the B complex, choline, is usually excluded from recommended Vitamin B supplements. It is also lamentable that choline is ignored in nutritional programs. Why it is left out of the minimum daily requirements chart is a mystery. Yet this part of the B complex is not only concerned with the body's health in general, but significantly associated with the function of the nervous system.

Choline is involved in every body function because it is required by the nervous system as a component of the complex nitrogenous hormone acetylcholine.

Acetylcholine is essential for the senses to report to the efferent nerves, for these nerves to report to other nerves, and for those to report to the brain. It is essential to the brain cells for thinking and for reporting back to the efferent nerves, and for these nerves to tell the muscles what to do.

At each point where the neurons or nerve cells report to each other there is a gap. This gap is called a synapse. The electrical nerve impulse cannot jump that gap. So the neuron releases from its axon (a sort of long telegraph wire) one of two hormones. It will release either epinephrine, which is used by the sympathetic or alarm nerve system and some of the parasympathetic nerve systems, or acetylcholine, which is used by the parasympathetic and central nervous system nerves.

Acetylcholine bridges the synaptic gap and carries the nerve message on, activating the receptor site on the dendrite or the cell body or the muscle fiber.

Another choline compound, the enzyme cholinesterase, breaks up and destroys the acetylcholine, so that the next message, a few thousandths of a second later, can get through.

Thus choline plays a leading role in transmitting the spark of the nerve impulse, especially to the voluntary skeletal muscles by the central nervous system and in the autonomic system in such important functions as cardiac inhibition. Without choline our hearts would stop beating.

WHEN YOUR NERVES NEED HELP

When you lack choline, the whole body can become weak and listless. Reputable doctors declare that a real deficiency can sometimes result in paralysis. Cardiac arrest—even death—is possible.

If it weren't for this nerve *spark*, you could not cause yourself to get out of bed in the mornings. It is only when your brain has the ability to transmit involved instructions to your arms and legs that your intentions are carried through the nerve fibers to the various muscles of your body.

It is then that you can go through the motions of bringing your body into a sitting position. Your body can be called upon to perform this one total movement with all of its muscular coordination only if you have provided the proper materials. In brief, you need to include choline in your daily diet. Perhaps the best way to assure yourself of an adequate supply is by eating lecithin at regular intervals.

YOUR LIVER ALSO REQUIRES CHOLINE

Now that you know that this essential nutrient, choline, has an indirect, but often critical, effect on nerve fibers, you are forewarned.

Because all fats must be metabolized in the liver in order for them to be properly utilized by the body, the liver must be kept in good working condition at all times.

In order to keep the liver in top working condition, it demands choline. Otherwise, fatty deposits have a way of building up inside that vital organ with the result that its hundreds of functions will be slowed down or blocked, thus throwing the whole body into a state of what we describe as poor health. This fact was proclaimed recently at an Atlantic City symposium of the American Institute of Nutrition. The results were reported in the January-February, 1971, Federation Proceedings.

Any of these untoward situations is less likely to occur when your diet includes the full maximum of choline with a minimum of fat intake.

Seven:

How to Milk a Soybean

Soy milk is one of the most unusual and nutritionally valuable forms of the soybean. To most persons, especially at first, the *milk* has a rather strong flavor of beans. However, proper handling can produce a product with a delightful nut-like taste.

There are many recipes for creating *soy milk* for human consumption—and endless claims have been made for its body-sustaining factors. However, when you get right down to cases, about the main advantages derive from the fact that soy milk is not mucous-forming, nor does it contribute to arteriosclerosis.

When the search for truth regarding soy milk was first started, it was quickly discovered that regardless of the directions you follow for preparing the milk substitute, the product can be of great value to persons who are allergic to animal milk.

Estimates vary as to the number of men, women and children in this country who have this problem, but as nearly as we can determine, one out of every five suffers an unfavorable reaction when cow's milk is used regularly or over long periods of time.

FOOD ALLERGIES CONTROLLED

On the other hand, soybeans, or soy milk have proved to be an excellent food for those with food sensitivities. When either item has been added to the diet, this versatile food tends to clear up pimples, rashes, eczemas, and other skin troubles. It has been noted, too, that the symptoms of arthritis and some forms of asthma have been greatly helped. The chief reason that soybeans are recommended for use in cases of eczema and other skin problems is that they render unnecessary the use of animal proteins such as meat, eggs, and milk, and thus lessen the chances of contracting inflammatory skin conditions. The soybean is apparently free from any of the common tendencies of allergic reactions which so frequently attend the use of all animal proteins.

ANOTHER SOURCE OF LECITHIN

One of the many virtues of the soybean is that it contains the essential lecithin. This by-product of the bean is a fat-soluble substance containing choline and phosphorus. Both are important aids to normal functioning of the body. And the lecithin content of the soybean seems to indicate that it is valuable in certain corrective diets. Discovery of the merits of lecithin is not new. In the beginning, it was extracted from egg yolks, but for many years the least expensive commercial source has been from soybeans. Normally, it is used in making candy and ice cream as well as bread and cooking oils. The purpose is to promote smoother blending and prevent rancidity.

LECITHIN, AN EMULSIFIER

In health products as an emulsifier, lecithin is supposed to break up fat and oil globules. This main function is to serve in spreading them evenly throughout the body to be disposed of as waste. In this connection, no one is yet precisely sure just how lecithin works in the human body. Research in major centers such as New York University, Northwestern University, and the University of Chicago, has indicated that among other things it promotes dispersal of deposits of fatty materials and cholesterol deposits in certain vital organs.

LECITHIN PLAYS IMPORTANT ROLE

Some scientists believe lecithin plays an important role in bringing these fats and fat-like substances under control. It is well to note that it has been established that lecithin is an important component of the myelin sheath around brain and nerve cells. Apparently this far-ranging substance is rich in factors which are important to the proper functioning of all living cells.

Many progressive doctors have suggested that lecithin might be used in the treatment of skin and nerve disorders.

LECITHIN HAS AN UNUSUAL HISTORY

The soybean has a distinguished history. It can lay claim to being one of the oldest crops cultivated by man. The first mention in Chinese records goes back nearly three thousand years. The soybean is closely associated with the history of China, where it has been an important source of protein for endless generations. Some historians boldly suggest the existence and survival of China as a nation is due to her use of soybeans as a staple food item.

Thus the cultivation of soybeans for a food crop has been known to man for fifty centuries. In matters of nutrition the peoples of the Orient were provided with many essentials that we obtain from other sources: milk, cheese, bread, oil and proteins. In trade, soybeans were a valuable crop that could be sold or exchanged for other necessities. In its original form, in the soybean, lecithin goes back a long way, in time and distance.

AMERICA LEARNS ABOUT THE SOYBEAN

Although Europe knew about this valuable legume almost a century earlier, the soybean was first introduced into the United States in 1804. Production in the United States did not reach an appreciable amount until 1924, when five million bushels were produced.

The increase in farm production of soybeans since that year has been little short of fantastic. At the beginning of World War II, it had reached the staggering figure of nearly eighty million bushels.

During the years of World War II, acreage devoted to the production of soybeans more than doubled. Some misguided people predicted that the soybean market had reached its peak and would fade out after the war. These gloomy prophets were wrong.

With modern research constantly finding new uses and creating improved methods of production, the increase has pushed steadily upward until today more than two hundred and forty million bushels are produced each year.

NEVER A SURPLUS

Demand has also grown in almost the same proportions. It is interesting to note that soybeans have never been classed as a surplus crop. Wheat, cotton, and other major crops have had to go begging to Congress for help even though production of soybeans has grown out of proportion to all expectations.

MORE VALUABLE THAN GOLD

The western world has now discovered that soybeans are more of an asset than gold in our modern civilization. During the last half-century soybeans have blossomed from a little-known forage crop into an agricultural phenomenon. The product has gained an important place in farming, business, food processing, and industry. Nutritionally soybeans filled a vital role in feeding a world at war. In industry the soybean is a challenge to the biochemist. At this time more than two hundred and fifty commercial products derive from the beans. Consequently, soybeans and the many products of soybeans, are destined to be a tremendous plus factor in our plans for tomorrow.

WE GROW IN KNOWLEDGE

We, in the United States, are rapidly learning the value of soybeans. Since we have only recently begun to use them as a food, we have not fully understood their nutritional merits. Nor have we acquired a taste for their savory flavor. We need to know more of their nutritional values and how to use them properly. Today it is no longer necessary for us to rely on the bean in its original form. Research has given us palatable soy products to meet every need. As time goes on, more and more people are discovering its unusual food values. Soybeans are growing in popularity every day.

Fortunately, the United States Department of Agriculture has done much to promote the development of the soybean industry. Early investigators commissioned by the government were sent to Asia to begin research on the soybean.

Soybeans can rightfully claim the distinction of

being one of the most nutritious foods known to man. They can help to balance a poor diet and add something new to make a good diet better. Vegetarians claim they can add zest and variety to their menus.

THE COW OF CHINA

When the price of milk goes up, most American mothers tremble. But not our more knowledgeable Chinese mothers. They would only smile disarmingly and put another handful of soybeans to soak. The reason for this unusual attitude is the fact that the soybean has long been the Cow of China. Children have been born and raised to adulthood in the Orient without knowing about any form of milk other than the beverage made from soybeans. It has served the people of China well—both as a food and as an item of trade value.

COST OF SOY MILK

Many interesting research projects on soy milk have been completed. One was based on costs for the first five months of an Oriental child's life. At the time the test was made fresh cow's milk in Shanghai cost twenty dollars a month in local money. At the same time prepared baby foods ranged in cost from twelve to sixteen dollars a month. But for the mother who fed her baby soybean milk, the slice off the family budget was only five dollars a month.

It is interesting to note that the *China Medical*

Journal, in the May issue of 1937, released test results of comparative values of soy milk, cow's milk and important baby foods. In a controlled study of several hundred children, the soy milk was rapidly digested and the children, ranging from birth to toddler age, showed firm symmetrical growth and notable muscular development. The report went on to stress the fact that for most children with digestive or malnutrition problems, the nutritional value of the protein contained in the soy milk was most helpful. When a child has not acquired the taste for cow's milk, he will drink soy milk and thrive.

SOY MILK HAS MANY USES

In content, soy milk is a colloid liquid. From this base, a lactic acid milk can be made. The next step is cheese. In China soybeans are not used so much as a vegetable as in the preparation of such dishes as cheese, sauce, breads, and meat substitutes.

When compared with cow's milk, soy milk is low in fat, carbohydrates, calcium, phosphorus and riboflavin, but high in iron, thiamine and niacin. For this reason, one or more of the following ingredients should be added to soy milk:

Honey—It helps to raise the carbohydrate content. It also improves the flavor.

Glucose, lactose, or maltose—preferable to honey if used for infant feeding.

Soy oil—it will increase the fat content. When properly blended should tend to give the consistency of cream.

Calcium Lactate—Should be added to all formulas, as needed to fortify them with calcium.

HOW TO MAKE SOY MILK

One simple formula for making soy milk in the home is explained in the following procedure: Start with one pound of a good grade of dried soybeans. Thoroughly wash and soak the beans in water overnight and store in a cool place, using two quarts of water for every pound of beans. Then pour off the water in which the beans have been soaking. Now grind the beans in your food mixer using the "chop" setting for fine grinding. As needed, add fresh pure water to the mixture. The amount of water you add should not be more than three times the volume of your soaked beans. When the mixing is fully completed, the whole batch is poured into a cheesecloth filter through which the liquid or resulting milk passes, leaving only the pulp. Save this residue. It has many uses in cooking. With constant stirring, the milk is heated, almost to boiling. Be careful not to scorch. Now skim off the surface froth. Cool the milk, but stir frequently to prevent the formation of Yuba or "soy milk skin".

Eight:

Health Seed of a Million Uses

From all indications it would seem that *diabetics* should have blood cholesterol determinations made by their doctor every few months. If this factor is excessive, every effort should be made to normalize the situation.

For example, it is known that the substitution of soy oil in place of saturated fats in diabetic diets has greatly reduced blood cholesterol in a matter of a few months. All foods needed to make use of fats and to help prevent clotting should be part of the daily intake of nutrients.

Lecithin and vitamin E are especially important, particularly when the threat of gangrene is present. Many times an amputation has been prevented simply by adding 600 units of vitamin E to the daily ration of food supplements.

Multiple sclerosis is characterized by calcified

patches on the brain and spinal cord. The particular traits of this disease show up in muscular weakness, poor coordination, and spasms of the arms and legs. Bladder control is also difficult.

Autopsies reveal a marked decrease in the lecithin content of the brain and a weakening of the myelin sheath covering the nerves. Both are normally high in lecithin content. In this connection even the lecithin is abnormal. Saturated fatty acids are present instead of the unsaturated type.

Multiple sclerosis is most common in nations of the world where the diet is particularly high in saturated fats. This invariably means that the amount of lecithin in the blood stream is greatly reduced. Marked improvement has been noted when three or more tablespoons of lecithin have been added to the daily intake of food.

It is generally felt that the lack of any nutritional value, whether it be choline, inositol, magnesium, vitamin B_6, or essential fatty acids, that prevents lecithin production can cause the condition of multiple sclerosis to worsen.

Apparently psoriasis, an eczema-like skin condition, is the result of faulty utilization of fats by the body. Persons with this skin problem usually have excessive amounts of cholesterol in their blood. When their blood cholesterol has been reduced to normal by eating plenty of lecithin the psoriasis has cleared up in a very short time. Persons with psoriasis were fed four to eight tablespoons of lecithin daily; new eruptions ceased within a few days. Severe cases recovered within a few months.

Psoriasis has also been helped by other vitamins

including A and B_6, but you should consult a competent doctor to determine how much.

Everybody should eat three tablespoons of granular lecithin daily with a balanced diet so as to help the liver produce its own lecithin.

Nephritis may result when a rapidly growing child is not given choline and he may be unable to obtain enough from methionine. When methionine, one of the essential amino acids, is radioactively labeled and injected into humans, it is recovered not as choline, but as lecithin. The basic cause of the fatty deposits that tend to damage the kidneys is apparently identical to that cause that induces arteriosclerosis.

In the condition known as Bright's disease, blood lecithin decreases in proportion to the seriousness of the ailment. When the blood cholesterol is extremely high, and the milky blood serum brought on by fatty substances is much in evidence, it is known that when lecithin is given these conditions are nearly always corrected.

When choline is added to an otherwise balanced diet, experimental nephritis is quickly brought under control.

IT IS CLAIMED LECITHIN RESTRAINS CLOTTING

Back in the gay nineties lecithin was represented as "the fuel that burns body fats." Blood fat is affected in the same way. Cholesterol reacts in the same manner. Lecithin causes the microscopic particles of fat in the blood of healthy persons to emulsify. In individuals subject to heart disease it causes the large molecules to separate into smaller bits which can then pass easily through arterial walls.

It is known that large fat particles apparently act as a foreign substance around which a clot may form. When this happens to blood circulation, cells tend to clump together and a clot begins to grow.

Men and women who have coronary problems, particularly young men, have invariably been found to be low in blood lecithin. When your lecithin is low, the danger of clotting is always present. Persons with a tendency towards heart attacks often have so much fat in their blood serum that it seems milky in appearance. When lecithin is eaten regularly this milky look quickly disappears.

In the diseased situation described as *arteriosclerosis*, any deficiency in the diet that prevents a normal amount of lecithin from being produced is dangerous. For example, when a lack of linoleic acid is present this deficiency indirectly allows blood fats to soar. Clots can then begin to form.

LECITHIN ADDED TO CALCIUM AND MAGNESIUM

Extensive studies in recent years point to the fact that almost everyone, especially persons in poor health, is deficient in magnesium. Unfortunately this is another mineral which has been largely sloughed off during the refining process. Furthermore, common farm practices coupled with the use of chemical fertilizers containing potassium prevent the more valuable magnesium from being absorbed by plants. Because of this short-sighted policy our foods are now extremely low in magnesium. Therefore it is more than likely that no other dietary deficiency is so much to blame for the widespread use of tranquilizers—or worse— drugs.

Our everyday diet should furnish us with approximately three hundred milligrams of magnesium daily, whereas five hundred to one thousand milligrams are apparently required for maintaining good health. This would seem to indicate supplementation is necessary. A tasteless magnesium oxide is available in 250 milligram tablets at your favorite health food store. However it is important that your magnesium intake should be approximately half that of the calcium you are eating.

Research has established the fact that when your calcium intake is excessive in relation to your magnesium intake, losses in urine can cause a magnesium deficiency. Obviously your magnesium intake should vary with the amount of calcium you are eating. When extra calcium is required, products are available which contain both calcium and magnesium in balanced proportions.

THE NEED FOR OIL IN REDUCING PROGRAMS

Since the major function of lecithin is to aid in burning off fats, the nutritional values needed to produce lecithin—linoleic or arachidonic acid, vita-min B_6, choline, inositol, and magnesium—are es-sential for a dependable reducing program. For example, when any oil is added to a diet lacking linoleic acid, a tremendous increase in energy is quickly in evidence. Too little linoleic acid has been known to damage the adrenals. When this happens the blood sugar is allowed to fall. Reduc-ing then becomes extremely difficult.

HAS THE NEED FOR LINOLENIC ACID BEEN PROVED?

Once again the adhesive tendency of blood cells known as platelets causes them to clump together and form a clot. Within a few hours this condition can be greatly reduced when a physician gives a coronary patient pure linolenic acid. Normally less than half a teaspoon. Soy oil rich in this essential fatty acid has proved to be helpful. It is known that one or two tablespoons of soy oil taken daily can prevent the tendency of the blood to clot dan-gerously. The linolenic acid treatment should be continued under the direction of your doctor in order to prevent the condition of abnormal ad-hesiveness of platelets from returning.

CAN LECITHIN BENEFITS BE ENHANCED?

When a condition we describe as good health prevails, it is possible to eat a meal high in fat or excessive in calories and the production of lecithin then increases greatly and the fat in the blood is promptly changed from large molecules to ever smaller and smaller ones. On the other hand, when a patient suffers with arteriosclerosis, blood lecithin stays disproportionately low regardless of the amount of fat entering the blood. With this problem fat molecules remain much too large to pass readily through arterial walls. A lack of lecithin in the cells may bring on a very precarious situation.

Cholesterol can be produced from fat, sugar, or indirectly from protein. Lecithin, however, has several parts. These parts require essential fatty acids and the B vitamins, choline and inositol, for their structure. Numerous other nutrients are needed to synthesize them. Because lecithin is essential to every cell in your body, the demand for these vitally important raw materials is enormous. An undersupply of any one of these factors limits the production of essential parts.

Fortunately, needed lecithin is available in the greatly preferred granular form which can be added to all foods with great benefit.

DOES LECITHIN HAVE AN EFFECT ON LIVER DAMAGE?

Without question a healthy liver is vitally important to our well-being. The incidence of liver damage, which includes the often fatal condition known as cirrhosis, is said to be increasing at an alarming rate. Even among young children. Formerly this problem was once confined to chronic alcoholics. It is now common with social drinkers, people who are overweight, and those who have been harmed by drugs or so called protective chemicals. Inadequate diets are also responsible.

Some doctors believe that the horrifying consumption of soft drinks—70 million bottles daily of one brand alone—is a major contributing cause. Obviously any toxic substance can harm the liver. Because all of us consume food additives, preservatives, nitrates from chemical fertilizers, DDT and other pesticides, and often water contaminated with detergent or water from reclaimed sewage, it is probable that everyone has suffered some liver damage. Good nutrition to prevent such damage should be our constant goal from this day forward.

The liver is our largest internal organ. Many hundreds of chemical reactions take place in the liver every minute. When the liver is working properly it deactivates hormones that are no longer needed. It synthesizes many important amino acids used in restoring body tissues. It breaks proteins into sugar and fat when this is required for physical energy or when we eat in excess of our needs. A healthy liver works hard to destroy harmful substances. It also detoxifies drugs and toxins from bacterial infections.

CAN LECITHIN BE USED FOR RECOVERY FROM LIVER DAMAGE?

People with damaged livers or even the affliction of cirrhosis can be helped to recover rapidly when their diets are corrected or improved. Even very bad cases have been known to recover within a few weeks. Normally improvement is much more rapid when lecithin is added to the daily intake of food.

HEALTH IN THE NUTRITION BANK

There are countless numbers of persons taking lecithin today who do not rush forth with flamboyant documentaries reflecting their personal results or feelings about the soybean derivative. These men and women are consistent in their consumption of either lecithin liquid, powder, wafers or granules, and would not elect to leave them out of their daily intake of food.

One states simply, "I started taking lecithin because of a high cholesterol count. I see improvement in my health."

Another, a self-diagnostician, shares these findings. "I have no way of determining its effect—just feel by what I've read and studied that it is a good emulsifier and helps prevent the accumulation of cholesterol."

Many of these persons intuitively feel a new mental alertness. Some see a change reflected in their children when lecithin granules are added to peanut butter, soups and salads to be shared by the whole family.

"I notice a change in my complexion," says an overweight man. "The constant flushed look is leaving, I think. I feel better too."

Relates another better informed woman, "I seem to feel an improvement since I switched from the lecithin liquid to the granules which also contain choline and inositol."

"My lecithin seems to do me more good when I include the B vitamins in my daily program," says another.

But even as these persons grope for better health and seek more knowledge and understanding, they are united with one thought. They like the soybean derivative of lecithin and what it seems to do for them. They also enjoy soybean foods and stock them on their pantry shelves.

Soy flour is not a flour in the sense of wheat or other grain flours. It is better described as highly concentrated vegetable food probably equal in food value to powdered eggs or dried milk solids.

The soybean food list is long and impressive. Its derivatives can be found in butter and lard substitutes, soy milk, salads, breakfast foods, infant foods, soy sauce, cakes, pastries, breads, bean curd, bean powder, canned vegetables, green vegetables, diabetic foods, macaroni, edible oils and salad oils, crackers and soy flour.

A VALUABLE HIDDEN FACTOR

Soybeans are valuable in many corrective diets. On the nutritional side we discover the fact that soybeans are rich in alkaline ash. This makes them high in essential potassium and other alkaline-bearing salts.

SOYBEANS PROVIDE A LOW-COST WHOLE FOOD

Soybeans supply an excellent low cost food item

that is nutritionally safe. They are one of our cheapest sources of essential protein. It is well to note that a few pennies worth of dry beans will serve four to six persons. Low income families can buy more food value in greater quantities in soybean products than in meat or fish products for the same price.

COMPARATIVE PROTEIN VALUES OF COMMON FOODS

When the composition of soy flour in vitamins and minerals is compared with the content of other flour, one can quickly judge their nutritional values.

Food	Grams Protein per Pound of Food
Buckwheat flour, light	29
Rye flour, light	40
Pecan meats	43
Wheat flour, patent	49
Rye flour, whole grain	50
Pork sausage	54
Whole wheat flour	59
Eggs	59
Frankfurters	68
Salmon	80
Lean Pork Chops	82
Halibut	86
Lean Beefsteak	90
Navy beans	100
American Cheese	108

Food	Grams Protein per Pound of Food
Dry whole milk	116
Peanut butter	118
Dry Milk Solids	160
Soya flour	
Low-fat	240
Medium-fat	225
Full-fat	180
Cottonseed flour	255
Peanut flour	265

THE BUTCHER, THE BAKER

Powdered soya is made from finely ground soya flour from which the soy oil has been removed. The oil is then refined and sprayed back into the flour. This method provides a finished product containing close to fourteen per cent refined oil. This mildly flavored product is preferred by bakers to the high-fat soya flour which contains a high percentage of oil. This stems from the fact that it has no shortening value.

Soya flour which has been lecithinated is a blend of soy flour, refined soy oil, and about five per cent of commercial lecithin, which is low in important choline and inositol factors. The added lecithin does have some advantages. It provides certain emulsifying and antioxidant properties which would not otherwise be present in soya flour. Soya flour and grits are gaining with meat packers as a valuable nutritional filler in sausages and wieners.

DO YOU NEED VEGETABLE OIL?

When you add any nutrient to your daily intake of food which is helpful in producing lecithin in normal amounts, you will find that it serves to slow down the condition of arteriosclerosis. Hydrogenated cooking fats, saturated animal fats, and nearly all low priced margarines contain little or no essential fatty acids. Consequently they cannot increase the production of lecithin. When vegetable oils rich in linoleic acids are added to the diet, low blood lecithin is elevated almost at once.

Moreover, when a solid fat, one of the five harmful items, is partly or fully replaced in controlled tests by soy oil then the blood cholesterol decreases as the body use of the oil improves. When the vegetable oil is gradually hydrogenated most tests reveal that the blood cholesterol goes up with each increase in hydrogenation.

Any oil, and this includes fish oils which contain no saturated fats, helps to reduce blood cholesterol. When vegetable oils rich in linoleic acid are added to the diet there is a marked increase in the quantity of cholesterol changed to bile salts. This steps up the breakdown of fats and cholesterol to carbon dioxide and water in the body tissues.

The amount of oil a person needs every day is apparently no more than one or two tablespoons. The more solid fats you eat, the greater becomes your need for linoleic acid. When your intake of solid fats is high, a corresponding deficiency in linoleic acid can be produced even though soy oil is included in your diet. In this connection it is well to know that there is nothing wrong with the natural saturated fats as long as your body cells are supplied with all the nutritional values required to make use of the intake of hard fats.

In practice, it is well to eat the same amount of fat as usual but decrease hard animal fats. All available evidence points to the fact that you should use oils for cooking, seasoning, and salad dressings. And by the same token avoid all hydrogenated margarines, cooking fats, and hydrogenated peanut butter. Processed cheeses and foods fall into this category.

It is urgently suggested that you do not go overboard in using soy oil. Normally each tablespoon supplies one hundred calories. When the extra calories are not used they begin to show up as soft, flabby fat. On this point I am sure your doctor will agree, oils alone cannot correct your problem of arteriosclerosis. Your diet still plays an important part in your health picture.

Nine:

Faith Without Action

There is a Biblical phrase that packs a powerful wallop: "Faith without action (works) is dead." When the full meaning of this basic truth is realized, you will be ready to accept the good that lecithin and its supporting substances can bring to you.

In this chapter you will find a chart that indicates possible *additives* that will bring tremendous power to the lecithin that you have included with your daily intake of food. The suggested uses that you can make of your *Better Living Tree*, are time-tested and valid. It remains for you to impart that very special reinforcement of *action*. The easy routines that are recommended should increase the beneficial forces of lecithin at least ten-fold.

To begin with, you should go to your nearest reliable health food store and purchase a bottle of lecithin wafers and a bag of lecithin granules. The wafers are convenient to eat with your meals and the granules may be added to your soups and/or salads.

You will note that there are sixteen branches on the *Better Living Tree*. Each arm of the tree has an important part to play in your good health pattern. Not a single one of the points should be neglected if you are sincerely determined to declare *"Mission Accomplished."*

PRIMARY BOOSTERS

Let's start with a discussion of the first directives:

AN URGENT SUGGESTION:
Be sure to have a regular checkup by your doctor —preferably one who is competent and sympathetic to your program.

ADDITIONAL BENEFITS:
Add one or two choline tablets or a teaspoonful of liquid inositol to the lecithin wafers which you eat each day.

A BALANCED INTAKE OF FOOD:
Precisely what this means is a matter to be determined by a well-informed nutritionist, or your family doctor. However, as a possible guideline it is suggested that you include some fresh fruit, vegetables, carbohydrates and protein with each meal—and *balanced* with enough vegetable oil to speed up the digestive processes.

DAILY DEEP BREATHING:
The power of breath cannot be minimized. The *tidal breath*, the natural minimum ebb and flow of air necessary to sustain life is not enough. Several times each day you should exhale all of the air from your lungs and then inhale deeply, retaining

the breath for a count of three. This simple exercise alone will triple the value of the lecithin that you consume.

PLENTY OF BED REST:

This means that you should relax completely for as many hours each day as your body requires for proper functioning, be it eight hours or five. This is something you will have to determine for yourself, but it is known that when your rest periods extend to seven or eight hours each day, your lecithin intake will be two to five times as effective.

POSITIVE THINKING:

This is a greatly overworked phrase, but it is so essential to your well-being that its value cannot be overestimated. In fact, when you permit negative thoughts to intrude into your consciousness, you *negate* nearly all of the benefits that you have achieved by adding lecithin to your diet.

WALK TWENTY BLOCKS EACH DAY:

Quite obviously, the very idea of walking twenty blocks every day will be a *shocker* to some people —*but it is a must*. The reason is that walking briskly for this distance speeds up the circulation, stimulates glandular activity and, best of all, helps to distribute the beneficial elements of lecithin to all parts of the body in need of repair.

DO NOT EAT ANY OF THE "FIVE WHITES":

What are the *"five whites"*? You are familiar with these:

White flour—lifeless

Refined white sugar

Hard fat (from meats, or worse, white lard)

Animal milk (I know this is a "*sacred cow*" but when the truth emerges, this highly-advertised food product will gradually fade out of our adult diet)

White rice

DAILY STRETCHING AND TENSING:

Once again, we need to activate all of our muscles as least once each day in order to accept the full benefits of lecithin. Isometric tension exercises serve this purpose *IF* this form of body movement is *limited to recommended use. NO MORE.*

BEGIN EACH DAY WITH A GLASS OF PURE WATER:

Why? Simply because water is an essential part of a properly functioning body—and, furthermore, water serves to carry the life-giving particles of lecithin to all parts of the anatomy. Consequently, an intake of six to eight glasses of water each day is another *must* that cannot be neglected.

SECONDARY BETTER LIVING BOOSTERS

Now we come to the final booster stage wherein we can grab on to and hold all of the marvelous advantages that are made available to us by partaking of some form of lecithin every day. To shirk the use of any one of the following powerful *additives* is to lessen the value of lecithin.

DAILY INTAKE OF VITAMINS:

Preferably, you should eat only the foods that pro-

Lecithin Better Living Tree

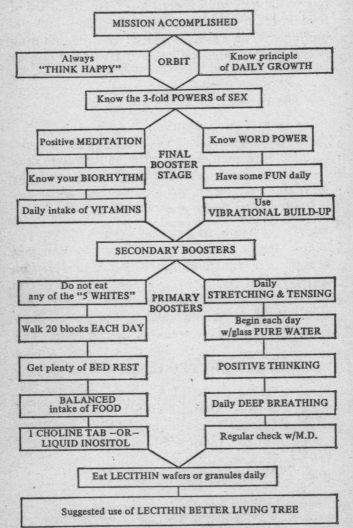

MISSION ACCOMPLISHED

Always "THINK HAPPY" — ORBIT — Know principle of DAILY GROWTH

Know the 3-fold POWERS of SEX

FINAL BOOSTER STAGE

Positive MEDITATION — Know WORD POWER

Know your BIORHYTHM — Have some FUN daily

Daily intake of VITAMINS — Use VIBRATIONAL BUILD-UP

SECONDARY BOOSTERS

PRIMARY BOOSTERS

Do not eat any of the "5 WHITES" — Daily STRETCHING & TENSING

Walk 20 blocks EACH DAY — Begin each day w/glass PURE WATER

Get plenty of BED REST — POSITIVE THINKING

BALANCED intake of FOOD — Daily DEEP BREATHING

1 CHOLINE TAB —OR— LIQUID INOSITOL — Regular check w/M.D.

Eat LECITHIN wafers or granules daily

Suggested use of LECITHIN BETTER LIVING TREE

vide enough of the natural vitamins and minerals, but when it is impossible to eat this quantity of food in order to take on your minimum daily requirements, then you should go to a reliable drug store, or a health food store that carries quality products and get a combination that is suitable to your particular needs. If there is any question, ask your doctor for his recommendation.

USE VIBRATIONAL BUILD-UP:

This tremendous benefit can be supplied mechanically, or brought on by mental stimulation through listening to good music, looking at beautiful pictures, or simply by repeating high-powered affirmations *with intensity* several times each day.

KNOW YOUR BIORHYTHM:

Let your personal Biorhythm chart guide you through the next twelve months. This can now be accomplished by computer by a Midwestern firm (Biorhythm-by-Computer, 6 North Michigan Ave., Chicago, Ill. 60602). With this valuable directional finder going for you, you can make your emotions work for you rather than against your way of life; and best of all, you will know your high and low energy days in three categories: Intellectual—Emotional—and Physical. This computer system has now been established as scientific fact. It is presently used by large industrial complexes and two major airlines. In daily use, your Biorhythm chart will help you to get your full share of benefits from the lecithin you eat.

HAVE SOME FUN EVERY DAY OF YOUR LIFE:

Fun is the catalyst that makes all of the other

guideposts to better living work to full capacity. Without fun, you will greatly lessen the inherent worth of lecithin. This vital, natural stimulant, fun, may be a game of cards, shuffleboard, golf, or just socializing harmoniously with other persons.

POSITIVE MEDITATION:

Most men and women regard meditation as simply sitting still and contemplating all "good." In truth, when you remain passive in your reflections you are getting nowhere. Positive meditation is quietly aggressive—indulged for a purpose. It can be health, success, or great accomplishment. Your affirmations should be, "I now have ... (whatever it is you want)" or, "I am now ... (Healed, restored to vibrant health, or immensely successful)" —and these meditative suggestions should be declared with strongly energized *intensity*. Then, and then only, will you gain outstanding benefits from the food that you eat.

KNOW THE POWER OF WORDS:

It is known that your physical ability to operate at top efficiency is tremendously enhanced when you repeat with *intensity* the twenty power-packed words that are supplied for your use in this chapter. In fact, it should increase the value of the lecithin that you eat *ten-fold*. This naturally goes for all of the food that you eat, but why not achieve the greatest possible benefit from the very special substance of lecithin that you include with your daily meals?

Power Words

USE THIS HIGHLY ENERGIZED
SOURCE OF POWER EVERY DAY

EACH WORD HAS A RESTORATIVE
BUILD-UP FACTOR OF FIVE

WITH TWENTY POWER-PACKED WORDS
YOU SHOULD RAISE YOUR POWER TO
ASSIMILATE FOOD VALUES TO 100%
WITHIN FIVE MINUTES.

HERE IS YOUR ENERGIZED WORD LIST

(1) Power	(8) Vitality	(15) Ecstasy
(2) Strength	(9) Health	(16) Pleasure
(3) Vigor	(10) Spirit	(17) Joy
(4) Energy	(11) Glow	(18) Charm
(5) Force	(12) Drive	(19) Thrilling
(6) Mighty	(13) Sparkling	(20) LOVE
(7) Strong	(14) Bold	

NOW GO INTO ORBIT

SEX REGENERATION:

Everyone is familiar with the procreative power of
sex, but very few persons know that there are two
more equally dynamic potentials involved in the
sex relationship. Unfortunately, most men and
women regard the second power—recreational sex
—as mere animal passion, when in truth it can be a
tremendously helpful experience, if it is used without fear of an unwanted conception. But the third
dynamic—*regeneration*—is all too often passed over

because it is like handling a sensitive bottle of nitroglycerine. This highly energized factor is un-clothed body contact—without indulgence. When this is used *with restraint*, physical powers of assimilation grow to fantastic proportions.

ALWAYS THINK HAPPY:

When you add any beneficial product to your daily intake of food, you can increase the value of the substance many times simply by *thinking happy* all during your waking hours. To succeed in the greatest venture of your life—*good health*—you must constantly practice the pose of happiness—*against all odds*.

KNOW THE PRINCIPLE OF DAILY GROWTH:

When you resort to eating vitamins and special mineral combinations with your meals, or include any special item, such as lecithin, you will be wasting your money unless you invoke the basic cause of *daily growth*. Back in the early twenties, one Emile Coué shook a complacent world to its very foundation when he created the phrase, "Day by day, in every way, I am getting better." Actually, this original thinker enunciated a fundamental of nature that must be used. There is no other way. The concept of daily growth—no matter how small the gain—is vitally essential to accepting the greatest good from the food that you eat. And this includes lecithin. You must consciously *know* that you are *improving* in health and well-being each and every day. Some call this psychosomatic, but in truth it is just plain common sense. You are now in orbit. Know that your mission is accomplished.